4th International Conference on Microbiome Engineering (ICME 2021)

AA008500

Online
3-5 November 2021

ISBN: 978-1-7138-4528-7

4th International Conference on Microbiome Engineering (ICME 2021)

Online
3-5 November 2021

AIChE
The Global Home of Chemical Engineers

Printed from e-media with permission by:

Curran Associates, Inc.
57 Morehouse Lane
Red Hook, NY 12571

CURRAN ASSOCIATES INC. proceedings .com

Some format issues inherent in the e-media version may also appear in this print version.

Copyright© (2021) by AIChE
All rights reserved.

Printed with permission by Curran Associates, Inc. (2022)

For permission requests, please contact AIChE
at the address below.

AIChE
120 Wall Street, FL 23
New York, NY 10005-4020

Phone: (800) 242-4363
Fax: (203) 775-5177

www.aiche.org

Additional copies of this publication are available from:

Curran Associates, Inc.
57 Morehouse Lane
Red Hook, NY 12571 USA
Phone: 845-758-0400
Fax: 845-758-2633
Email: curran@proceedings.com
Web: www.proceedings.com

TABLE OF CONTENTS

Automated and Machine Learning Guided Culturomics of Personalized Gut Microbiomes

Authors:

Huang, Y. - Presenter, Columbia University

Wang, H., Columbia University

Sheth, R. U., Columbia University

The human gut microbiome is made up of hundreds of strains, exhibiting extensive interpersonal diversity. Current approaches to build gut strain biobanks to isolate and study individual strains are labor-intensive, not scalable, and lack integration of strain phenotypes with genomic data. Here, we describe a high-throughput system that leverages phenotypic and genomic data for automated anaerobic strain isolation and cultivation. This system was used to establish an extensive and personalized collection of gut strains from 12 healthy individuals composed of 16,584 isolates and 653 high-quality whole genomes spanning 358 unique taxa, which in aggregate covered >80% of the human gut microbiome. We demonstrated that the morphology and spatial arrangement of colonies can predict taxonomic identity, species-specific growth profiles, and interspecies interactions. Comparative analysis of intra- and inter-personal strain-level genomic variations revealed patterns of gut microbiome evolution, selection, and horizontal gene transfer. Automated platforms that link taxa-level phenotypic data with genomic data will serve as a key resource to build personalized microbiome biobanks on-demand and help unlock the mechanistic basis of gut microbiota in health and disease.

Commensal Resilience Mechanisms in the Inflamed Intestine

Authors:

Zhu, W. - Presenter, Vanderbilt University Medical Center

Chazin, W., Vanderbilt University

The gut microbial community is constantly subjected to perturbations such as intestinal inflammation, which could drive the microbiota into a perturbed state termed dysbiosis. Research in the microbiome has primarily focused on the factors that induce changes in the gut microbial community, while little is known about how the microbiota responds to perturbations and remains resilient. A hallmark change that occurs during episodes of intestinal inflammation is the marked reduction of iron bioavailability, a consequence of a series of host processes known as nutritional immunity. Pathogenic bacterial such as *Salmonella* Typhimurium (*S.* Tm) have evolved to produce small, high-affinity iron-chelating molecules termed siderophores to overcome host nutritional immunity and thrive in the inflamed intestine. Here, we show that commensal gut *Bacteroides thetaiotaomicron* (*B. theta*) acquires iron in the inflamed gut by pirating siderophores from an enteric pathogen that causes intestinal iron limitation. Notably, *B. theta* captures siderophores using a unique system absent in other Gram-negative bacteria. However, such a capture mechanism can be exploited by enteric pathogens to â€œre-pirateâ€� siderophores from gut commensals to evade nutritional immunity, adding a previously unappreciated dimension to the intricate interactions between pathogen and nutritional immunity.

Engineering a Gut Biosensor for the Detection of Colitis Using the Probiotic Yeast Strain Saccharomyces Boulardii

Authors:

Dacquay, L. - Presenter

McMillen, D. R., University of Toronto

Philpott, D. J., University of Toronto

Tsang, D., University of Toronto

Inflammatory bowel disease presents a diagnostic challenge in that periodically monitoring inflammation within the intestine is difficult without invasive methods. A more convenient method to detect intestinal inflammation would be helpful to confirm clinical diagnoses and to evaluate the efficacy of therapeutic approaches. There is increased interest in using probiotics for *in situ* detection of intestinal diseases since they are able to survive passage through the intestine providing an opportunity to design genetically encoded biosensors to detect disease biomarkers. Here we describe progress towards an intestinal inflammation biosensor in the probiotic yeast, *Saccharomyces boulardii*. We recently engineered an oxidative stress sensing biosensor within *S. boulardii* that has a very strong and sensitive response towards reactive oxygen species- a common marker of inflammation. Based on this sensor, we are developing a probiotic yeast reporter that could be ingested by a patient, recovered in a stool sample, plated, and visually inspected to see if it had encountered inflammation. To this end, we modified a Cre/lox system to switch the characteristic colony colour of *S. boulardii* from white to red upon detection of reactive oxygen species through the excision of a genomically integrated lox-flanked *ADE2* cassette; the final output is then the fraction of plated red and white colonies. Here, we present results on the optimization and tuning of the oxidative stress sensor, along with preliminary results from a DSS mouse model of colitis. We believe that the results of our work could prove useful for the development of other probiotic yeast-based biosensors.

Engineering Phages for Eukaryotic Payload Expression in Cancer Associated Fusobacterium Nucleatum

Authors:

Zelcbuch, L. - Presenter, BiomX

Zigelman, Y., BiomX

Kowalsman, N., BiomX

Ninio-Many, L., biomX

Buchshtab, N., BiomX

Vainberg Slutskin, I., BiomX

Elharar, Y., BiomX

Safyon-Gartman, E., BiomX

Weiner, I., BiomX

Berko-Ashur, D., BiomX

Nicenboim, J., BiomX

Bassan, M., BiomX

Fusobacterium nucleatum (*FN*) species are enriched in human colorectal tumors [1] and their presence is correlated with poor prognosis [2]. Bacteriophage ("phage") have high intrinsic safety and specificity and offer a promising treatment strategy to target *FN* associated with colorectal cancer (CRC). Phage engineering allows specific delivery of anti-tumor immune stimulating payloads. A major obstacle in phage engineering is to express eukaryotic payloads in unconventional prokaryotic bacterial hosts, such as *FN*, while ensuring high expression and activity.

To address these challenges, a *de novo* molecular system was established. Initially, a shuttle vector was generated using a native *FN* plasmid. Two promoters that drive strong bacterial gene expression in *FN* were identified using both bioinformatic tools and mass spectrometry and their activity was validated using a luciferase-based detection tag.

Payload protein expression was further improved using various algorithms for exposing the ribosome binding site and optimizing codon usage and ribosome elongation speed. After determining the optimal conditions for payload expression via a plasmid, a *FN* phage was engineered to carry the payload within its genome. Following phage infection, payload expression by the bacterial host was observed.

Expanding the Therapeutic Capacity of Engineered Commensal Microbes

Authors:

Kelly, V. - Presenter

Sirk, S., University of Illinois at Urbana Champaign

One of the great challenges in the development of biotherapeutics is the high cost associated with these drugs, both in terms of manufacturing cost and the patientâ€™s cost for treatment. Dosing regimens requiring repeated administration, trips to a healthcare provider, and large amounts of formulated therapeutic, are a significant driver of these costs. The goal of our research is to leverage the inherent capabilities of the diverse microbial community of the human gut to reduce these costs by acting as therapeutic factories *in situ*. Recent developments in microbial genetics, synthetic biology, and the study of niche-optimized, non-pathogenic members of the human gut microbiota have paved the way for the engineering of commensal bacteria to carry out novel functions. Using species-specific genetic components, therapeutic production can be achieved in both transiently and persistently colonizing members of the gut microbiota. While these engineered strains can be used to treat conditions localized to the gastrointestinal tract, we are currently utilizing a protein engineering approach to expand the potential applications of engineered microbes to systemic conditions. We have fused small (15 amino acid) peptides that mimic the neonatal Fc receptor (FcRn)-binding epitopes of IgG and albumin to single-chain Fv antibody fragments. These modifications enable pH-dependent FcRn engagement and FcRn-mediated salvage or transport across polarized epithelial cell barriers, functionally mimicking the native FcRn-mediated half-life extension and transport of IgG and albumin. This work demonstrates the potential utility of peptide-modified scFvs either as purified therapeutics or as part of an *in situ* engineered microbial delivery system.

Microbiome Interaction Network Toplologies to Improve Probiotic Ingress, Colonization, and Persistence across Varied Resident Communities

Authors:

Sander, K. - Presenter, University of Tennessee/Oak Ridge National Laboratory

Sharpless, W., University of California, San Diego

Friedline, S., University of British Columbia

Arkin, A. P., University of California, Berkeley

There is an increasing interest in delivering microbial probiotics to support the health of plants and animals. However, a challenge remains in ensuring that the active members of the probiotic can engraft into the host community sufficiently to deliver their desired effect. While there are many factors that can effect whether an organism will survive in a given environment, one of the critical factors is competition with the endogenous community members already present. The sporadic and unexplained failure of foreign species to inhabit environments suggests a need to consider resident microbial communities and their interspecies interactions when designing synthetic microbial community probiotics.

We hypothesize that a probiotic community can be composited such that supportive interactions among its members greatly increase the chances of survival and growth of its active members regardless of variation in composition of the endogenous population. We present a modeling and design framework to test this hypothesis and identify candidate synthetic community interaction networks which may serve this purpose.

Our microbial network optimization framework yielded microbial interaction network motifs for synthetic probiotics which, in simulation, promote better engraftment of a target species into diverse resident communities. Identified probiotic interaction networks are enriched for exploitative interactions between probiotic species. We find motif sign topology is critically important, while interaction magnitudes and species growth rate have little effect when varied within biological bounds. These results offer interaction motif templates that can be used to design synthetic microbial communities which reliably engraft into existing environments and their resident communities.

Multi-Input CRISPR-Based Kill-Switches for Biocontainment of Engineered Microbes

Authors:

Moon, T. S. - Presenter, Washington University in St. Louis

Rottinghaus, A., Washington University In St. Louis

Ferreiro, A., Washington University School of Medicine

Dantas, G., Washington University in Saint Louis

Probiotics are effective chassis for diagnostic and therapeutic applications. However, there are safety concerns associated with using genetically engineered organisms for medical applications. To address these issues, we engineered *Escherichia coli* Nissle to survive only when and where it is needed using CRISPR-based kill-switches (CRISPRks). We first designed a CRISPRks that induces cell death by expressing Cas9 and genome-targeting guide RNAs (gRNAs) in response to a chemical inducer. This design allows cell killing to occur while the microbe is in the gut in response to oral administration of the chemical. We optimized the efficiency and stability of the CRISPRks, achieved more than a 9-log reduction in cell number, and demonstrated genetic stability for up to 28 days of continuous growth. This high killing efficiency was maintained *in vivo*, where we achieved complete elimination of the probiotic after oral administration of the inducer. To our knowledge, this is the first time on-demand elimination of an engineered microbe that has been demonstrated *in vivo*. We next modified our chemically inducible-CRISPRks to also induce cell death in response to ambient temperatures below 33â€™C. This design induces cell killing either in response to oral administration of the chemical or when the microbe is excreted from the body in response to the reduced environmental temperature. Our CRISPRks strategy provides a benchmark for future microbial biocontainment circuits. The sensor and killing mechanism are well characterized and functional in many microbes, allowing the CRISPRks design to be broadly applicable in diverse microbiota.

Roma's Abstract, please ignore for now

Authors:

Alexander, L. - Presenter, AIChE

Homologous recombination combined with CRISPR-Cas counterselection is a powerful approach for genome editing in bacterial species. However, genome editing efficacy is dependent on host factors and (the efficiency of) DNA repair pathways of the host. To mitigate these constrains, we developed **SIBR-Cas** (Self-splicing Intron Based Riboswitch-Cas), which functions as a modular, tightly regulated and inducible system that can control the expression of *FnCas12a* in a wide range of bacterial species. We chose the Group I T4 bacteriophage *thymidylate synthase* (*td*) intron as the appropriate mechanism to control *FnCas12a* and repurposed it into a universal and inducible genome editing system. SIBR-Cas was applied successfully to knock-out several genes in three distinct wild type bacterial species (*Escherichia coli*, *Pseudomonas putida* and *Flavobacterium* IR1) with poor homologous recombination systems, without the use of exogenous recombinases nor use of inducible promoters. Compared to other genome engineering tools, SIBR-Cas is simple (one plasmid and independent of host factors), tight (little to no leakage), inducible (only active when inducer is applied in the medium) and widely applicable (can be applied to virtually any gene/RNA of interest and any bacterium under any promoter), allowing for the full utilization of non-model bacteria. Finally, we propose that SIBR-X (X stands for any gene/RNA of interest) can have a wider application as a universal gene expression control mechanism for virtually any gene of interest in prokaryotic organisms.

Sibr-Cas Enables Host-Independent and Universal CRISPR Genome Engineering in Prokaryotes

Authors:

Patinios, C. - Presenter, Wageningen University and Research

van der Oost, J., Wageningen UR

Arifah, A., Wageningen University and Research

Perez, B., Wageningen University and Research

Staals, R., Wageningen University and Research

Kengen, S., Wageningen UR

Ingham, C., Hoekmine BV

Creutzburg, S., Wageningen University and Research

Homologous recombination combined with CRISPR-Cas counterselection is a powerful approach for genome editing in bacterial species. However, genome editing efficacy is dependent on host factors and (the efficiency of) DNA repair pathways of the host. To mitigate these constrains, we developed **SIBR-Cas** (Self-splicing Intron Based Riboswitch-Cas), which functions as a modular, tightly regulated and inducible system that can control the expression of *FnCas12a* in a wide range of bacterial species. We chose the Group I T4 bacteriophage *thymidylate synthase* (*td*) intron as the appropriate mechanism to control *FnCas12a* and repurposed it into a universal and inducible genome editing system. SIBR-Cas was applied successfully to knock-out several genes in three distinct wild type bacterial species (*Escherichia coli*, *Pseudomonas putida* and *Flavobacterium* IR1) with poor homologous recombination systems, without the use of exogenous recombinases nor use of inducible promoters. Compared to other genome engineering tools, SIBR-Cas is simple (one plasmid and independent of host factors), tight (little to no leakage), inducible (only active when inducer is applied in the medium) and widely applicable (can be applied to virtually any gene/RNA of interest and any bacterium under any promoter), allowing for the full utilization of non-model bacteria. Finally, we propose that SIBR-X (X stands for any gene/RNA of interest) can have a wider application as a universal gene expression control mechanism for virtually any gene of interest in prokaryotic organisms.

Single-Cell Fusion PCR for Biocontainment Surveillance

Authors:

Diebold, P. - Presenter, Cornell University

Engineered microorganisms hold great potential for shaping complex industrial and environmental processes. However, introducing genetically modified organisms into an environment always holds the potential for horizonal gene transfer, thus requiring strong biocontainment strategies. While significant energy has been directed towards biocontainment, it is currently difficult to assess the effectiveness of these methods. Common molecular methods such as short- and long-read sequencing of microbiomes cannot always identify the hosts of rare or extrachromosomal genes. One-step Isolation and Lysis PCR (OIL-PCR), which molecularly links any gene of interest with the bacterial 16S rRNA gene via fusion PCR performed within an emulsion, is an ideal method to assay the effectiveness of biocontainment strategies. OIL-PCR is highly sensitive and has been validated to accurately associate plasmid encoded genes with the bacterial hosts in a complex stool community with up to 99% accuracy.

The Mystery of Built Environment Microbiome : Mapping Bacterial Diversity and Translationally Active Abundance in Three Different Office Buildings

Authors:

Singh, J. - Presenter, University of Delaware

Blenner, M., University of Delaware

Monteiro, L., University of Delaware

The microbiota of the built environment is a combination of human and environmental sources. HEPA filters as an ecological niche for indoor microorganisms are critically important for assessing indoor environmental health. Herein, we used 16S rRNA amplicon sequencing to analyze the bacterial communities in HEPA filters of three office buildings. Chao1, Shannon, and phylogenetic diversity values in the filter of two buildings experiencing the most human interaction was greater than that of the third building experiencing the least interaction. The bacterial source composition varied among buildings; Alphaproteobacteria, Actinobacteria, Bacteroidia, and Bacilli were the predominant. A larger proportion of the bacteria in the HEPA filters of the first two buildings were attributed due to human occupancy. The soil, however, contributed to a higher proportion of the bacteria in the HEPA filters of all buildings. Despite the lack of direct human interaction, filter samples from building 3 contained 15% to 25% human skin-associated taxa. Seven genera were found as the core microbiota of all the buildings, of which *Acinetobacter*, *Enterobacter*, *Pseudomonas*, and *Staphylococcus* are associated with human health. Additionally, BONCAT-FACS was used to determine the translationally active bacterial population in three buildings. It can detect active cells by tracking the incorporation of amino acids into newly synthesized proteins through bioorthogonal noncanonical amino acid tagging (BONCAT). Approximately 30 to 65 percent of cells of the filter microbiome in all buildings were found to be translationally active. Overall, this study demonstrates sources of indoor contamination in three office buildings and their metabolic state.

Characterisation and Engineering of Probiotic Saccharomyces Boulardii Promoters for Robust Therapeutic Production In Vivo.

Authors:

Sands, C. - Presenter

V. Uribe, R., Technical University of Denmark (DTU)

Sommer, M. O. A., Technical University of Denmark

Robust expression of therapeutics from advanced microbial therapeutics is dependent on understanding the expression profiles of relevant promoters in the gastrointestinal (GI) tract. *Saccharomyces boulardii* is a probiotic yeast that has been used to produce therapeutic molecules and proteins in the GI tract. While promoters of *S. boulardii* have been characterised under standard laboratory conditions, their activity *in vivo* is not known. In this study we characterise a selection of relevant promoters *in vitro* under gut-like conditions (including microaerobic and anaerobic conditions, and alternative carbon sources), to identify candidate promoters for further optimization. Next, promising promoters will be engineered to generate a library of robust synthetic promoters with standard expression profiles. Finally, the engineered promoters will be validated *in vivo* with the production of a therapeutic molecule. The study will provide researchers working with *S. boulardii* a comprehensive dataset on the expression profiles of commonly used native promoters, as well as providing a library of engineered promoters, validated *in vivo*.

Top-Down and Bottom-up Approaches for Investigating the Microcystis Phycosphere

Authors:

Tan, J. - Presenter, University of Michigan

Smith, D., University of Michigan

Rivera, S., University of Michigan

Lin, X., University of Michigan-Ann Arbor

Dick, G., University of Michigan

Understanding how cell-cell interactions compound into dynamics observed in natural communities is crucial for predicting and engineering microbial communities. Both metagenomic studies of natural communities and â€œsynthetic ecologyâ€ with defined microbial communities have provided insight into composition, dynamics, and potential mechanisms. However, for microbial systems that lack well-characterized and tractable representatives for synthetic ecology, they are heavily limited to study by culture-independent techniques. Despite tremendous improvements in culture-independent techniques, the high-resolution needed to study these interactions in natural communities is confounded by their sheer complexity. This work presents two approaches, one top-down approach and one bottom-up approach, to provide insight into interactions between difficult-to-study, toxic bloom-forming *Microcystis aeruginosa* and the heterotrophic partners that inhabit the carbohydrate-rich matrix surrounding *Microcystis* cells - referred to as the phycosphere. The top-bottom approach involves using droplets to encapsulate single *Microcystis*-heterotroph colonies from blooms in Lake Erie to study membership and strain-level association at the local scale. The bottom-up approach utilizes microfluidic droplets, which are sub-nanoliter to microliter scale water-in-oil emulsions. These â€œmicrodropletsâ€ serve as miniaturized, self-contained bioreactors for microbial cultivation, rendering ultra-high throughput investigation. In particular, the small size and stochastic nature of microdroplets enable the decomposition of a mixed culture composed of *Microcystis aeruginosa* and heterotrophic partners to interrogate critical pairwise interactions. We believe that both efforts can bridge the mechanistic understanding from synthetic ecology with the observed ecology from metagenomic studies to understand how the microbial partnerships contribute to the resiliency of *Microcystis*-dominated harmful cyanobacterial blooms.

AUTHOR INDEX

Notes:

Notes:

Notes:

Notes:

AIChE
120 Wall Street, FL 23
New York, NY 10005-4020

ISBN 978-1-7138-4528-7

CURRAN ASSOCIATES INC.
proceedings.com